PHILIP'S

GUIDE TO THE
NIGHT SKY

A GUIDED TOUR OF THE STARS & CONSTELLATIONS

PHILIP'S

GUIDE TO THE
NIGHT SKY

A GUIDED TOUR OF THE STARS & CONSTELLATIONS

SIR PATRICK MOORE

SIR PATRICK MOORE is the author of more than 100 books, and presenter of the world's longest-running television series, *The Sky at Night*. He is a Fellow of the Royal Astronomical Society (Jackson-Gwilt medalist), a member of the International Astronomical Union, and holder of the Goodacre Medal from the British Astronomical Association, of which he was president from 1982–84 and is now Life Vice-President. Sir Patrick is also a member of astronomical societies of many other countries.

In 1967 he was awarded the OBE for his services to astronomy, in 1988 the CBE, and in 2001 he received a knighthood. A minor planet (No. 2602) has been named Moore after him. Sir Patrick holds honorary doctorates from several universities, including Lancaster, Birmingham, Hertfordshire, Keele, Leicester, Portsmouth, Glamorgan, Sheffield Hallam and Trinity College Dublin.

First published in 2005 by Philip's,
a division of Octopus Publishing Group Limited
(www.octopusbooks.co.uk)
Endeavour House, 189 Shaftesbury Avenue,
London WC2H 8JY
An Hachette UK Company (www.hachette.co.uk)

This new edition published 2005
Reprinted 2007, 2008, 2009, 2010, 2011

ISBN 978–0–540–08701–3

Printed in China

Details of other Philip's titles can be found on our website at: **www.philips-maps.co.uk**

Title page: An infra-red view of the Eagle Nebula obtained by the Very Large Telescope.

CONTENTS

CHAPTER ONE

THE STARLIT SKY

Have you ever tried to find your way around the night sky? It may seem a daunting task, but in fact there is nothing at all difficult about it, and the stars become so much more interesting when you know which is which. It is not true that on a clear night you can see 'millions of stars', as is often thought. If you can see two and a half thousand stars, you are doing very well indeed.

The star-patterns or *constellations* do not change in position over periods of many lifetimes, so once you have identified a constellation it will always look the same. Of course, it will not be in the same position in the sky, because the Earth spins round from west to east, causing the whole sky to turn from east to west in a period of 24 hours and taking the Sun, Moon and stars with it; but the familiar patterns are unaltered.

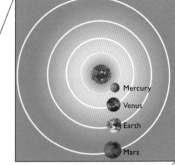

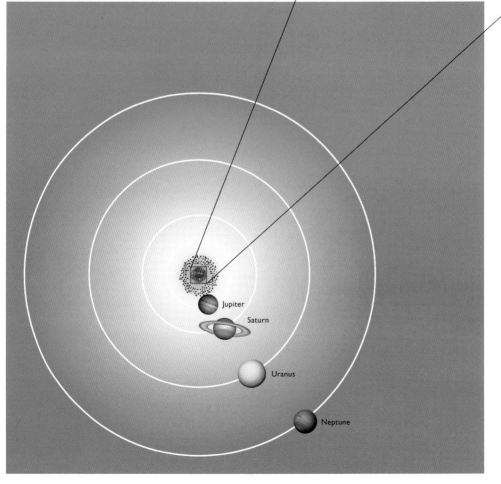

THE EARTH, SUN AND MOON

The Earth is a *planet*, moving round the Sun at a distance of 150 million kilometres (93 million miles). There are seven other planets, moving round the Sun at different distances and in different times: Mercury and Venus are closer to the Sun than we are, while Mars, Jupiter, Saturn, Uranus, and Neptune are further away. Together with various bodies of lesser importance, such as the Kuiper Belt object Pluto, and the moons or *satellites* of the planets, these make up the Sun's family or *Solar System* – our own particular part of the universe. The planets have no light of their own, and shine merely because they reflect the rays of the Sun. They look like stars as seen with the naked eye, but, unlike the stars, they wander slowly around from one constellation to another, even though they always keep to certain well-defined regions. Mercury always stays so close to the Sun that it is not easy to find unless you look deliberately for it at the best possible moment; Uranus is on the fringe of naked-eye visibility, and to see Neptune you need optical aid. The other planets are bright, but it is possible to confuse them with stars unless you know where they are.

The Sun is an ordinary star. It is a globe of hot gas, with a surface temperature of 6000°C, and it is big enough to swallow up more than a million globes the volume of the Earth. It is shining not because it is burning in the manner of a coal fire, but because of actions going on deep inside its globe, where the temperature rises to the staggering value of 14 million degrees C and perhaps rather more. The Sun contains a great deal of hydrogen, which is the most plentiful substance in the entire universe (after all, water is made up partly of hydrogen – the famous formula H_2O means that a water molecule is composed of two atoms of hydrogen together with one atom of oxygen). Near the Sun's core, the hydrogen is being changed into another gas, helium. It takes four 'bits' of hydrogen to make one 'bit' of helium; every time this happens, a little energy is set free and a little mass is lost. This is how the Sun shines. The loss in mass (or 'weight') amounts to 4 million tons every second, so the Sun weighs much less now than it did when you opened this book, but there is no need for alarm; the Sun will not change much for at least a thousand million years in the future.

I must give a warning here. Because the Sun is so bright and hot, it is unwise to stare straight at it – and at all costs, never look at the Sun through any telescope or binoculars, even with the addition of a dark filter. Permanent damage to the eye is certain to result. As I have said many times, there is only one rule about looking straight at the Sun through a telescope: **don't**.

The Moon, our satellite, is quite harmless, because it sends us almost no heat. It is the Earth's faithful companion, and stays together with us as we journey round the Sun; its distance from us is on average 385,000 kilometres (239,000 miles), which is less

◄ *Plan of the Solar System. The orbits of the various planets; the four inner planets are shown to a larger scale in the inset, upper right. Not to scale.*

◄ *Earth and Moon compared. The Moon's diameter is little more than a quarter that of the Earth.*

than ten times the distance right round the Earth's equator. Like the planets, it shines by reflected sunlight. The *phases*, or monthly changes of shape from new to full, depend upon how much of the Moon's sunlit side is turned in our direction.

The Moon is smaller than the Earth, with a diameter of only 3476 kilometres (2160 miles) as against 12,756 kilometres (7926 miles) for the Earth. (Represent the Earth by a tennis ball, and the Moon will be about the size of a table-tennis ball.) It has a lower pull of gravity, and so it has been unable to cling on to any atmosphere it may once have had, so today it is airless and lifeless; the broad dark plains, easily visible with the naked eye, are still called 'seas', but there has never been any water in them. The whole of the Moon is dominated by mountains, valleys and craters. Binoculars show them splendidly, and the Moon is a fascinating world, but one has to admit that near the time of Full Moon the stars are drowned in the brilliant light, so only the main constellations can be seen.

THE STARS

All the stars visible at night-time are themselves suns. They look so small only because they are so far away from us. In fact, they are so remote that it would be clumsy to give their distances in inches or centimetres (just as it would be awkward to give the distance between London and Edinburgh in feet or metres), and astronomers use a different unit, the *light-year*. Light travels at 300,000 kilometres (186,000 miles) per second; in a year it cov-

► *The Half Moon. This is the view of the 'last quarter' Moon, after full. The Apennine Mountains are seen to the upper right, forming part of the boundary of the lunar 'Sea of Showers'.*

ers almost 9.7 million million kilometres (6 million million miles) and this is the astronomical light-year. Light takes one and a quarter seconds to reach us from the Moon, and only 8.6 minutes from the Sun, but the nearest star beyond the Sun is over four light-years away.

This immense distance scale explains why the stars remain in the same relative positions, or virtually so, for century after century; the constellations we see today are to all intents and purposes the same as those which were seen by Julius Caesar. The stars are not genuinely fixed, and are moving around in all sorts of directions at all sorts of speeds, but their individual or *proper* motions seem very slight. Consider first a bird flying among the tree-tops, and then a jet aircraft against the clouds. The bird will seem to be moving the faster of the

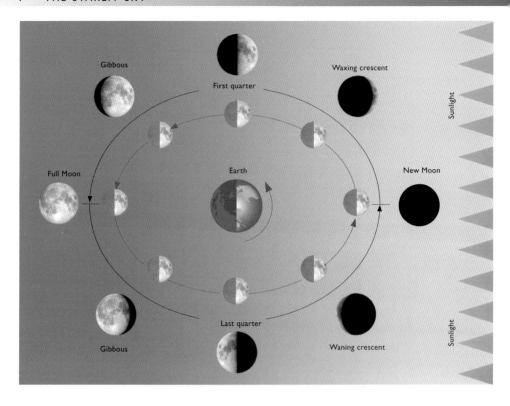

Gibbous

Waxing crescent

First quarter

Sunlight

Full Moon

Earth

New Moon

Last quarter

Gibbous

Waning crescent

Sunlight

two, because it is closer, but I doubt any bird has yet broken the sound barrier! The stars are so far away that they seem to 'stop'.

▲ *Phases of the Moon. Not to scale.*

The daily rotation of the sky is, as we have seen, due to the real rotation of the Earth on its axis. Northward, the axis points to the *celestial pole*, which is marked fairly closely by a brightish star called Polaris, or the Pole Star, in the constellation of Ursa Minor (the Little Bear). This is why Polaris seems to remain almost motionless, with everything else moving round it. It does not mean that Polaris is itself of special importance, even though it is much larger and more luminous than the Sun; it simply happens to lie in this particular position in the sky.

Stars near the celestial pole will move round it in circles, never dropping below the horizon; these are called *circumpolar* stars. Ursa Major (the Great Bear) is circumpolar from Britain. Stars further away from the pole spend part of their 24-hour circulation below the horizon, so they cannot always be seen; such is Arcturus, the brilliant orange star in Boötes (the Herdsman).

The view of the sky depends upon your position on the surface of the Earth. From Britain, we can never see some of the splendid groups of the far south of the sky, because they never rise above our horizon, while from New Zealand it is never possible to see the Pole Star or the Great Bear. The maps in this book have been drawn for the latitude of London, but they can be used anywhere in the British Isles, northern Europe or northern North America.

The stars are very different in brightness and colour. We measure the apparent brightness of a star by its *magnitude*. The scale works rather in the manner of a golfer's handicap, with the more brilliant performers having the lower values. Thus a star of magnitude 1 is very bright; 2, less bright, and so on down to magnitude 6. To see stars below magnitude 6, you need binoculars or a telescope. Most of the stars shown in the maps in this book are of magnitudes 1, 2 or 3, so they stand out.

The colour of a star depends upon its surface temperature. Our Sun is yellow. A star with a cooler surface will be orange or red, while a hotter star will be white or bluish. These different colours indicate different stages in a star's life-cycle; the red stars are older than the hotter ones. Our yellow Sun is just about middle-aged.

It is important to realize that our Sun is unexceptional. We know of stars which are much more powerful. The Pole Star is 6000 times as luminous as the Sun; Rigel, in Orion, could match 60,000 Suns, but is over 900 light-years away. Look at Rigel tonight, and you will see it as it used to be in the time of William the Conqueror. Once we look beyond the Solar System, our view of the universe is bound to be very out of date.

▼ *Circumpolar and non-circumpolar Stars. Ursa Major (the Plough) never sets as seen from Britain, but Arcturus does.*

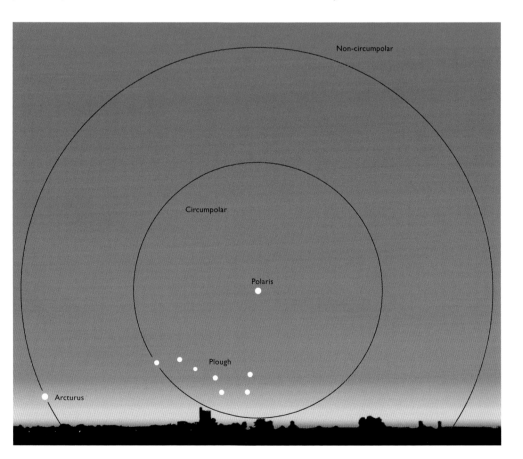

Non-circumpolar

Circumpolar

Polaris

Plough

Arcturus

▶ *Shape of our Galaxy. The Galaxy is a flattened system; when we look along the main plane, we see the appearance of the Milky Way.*

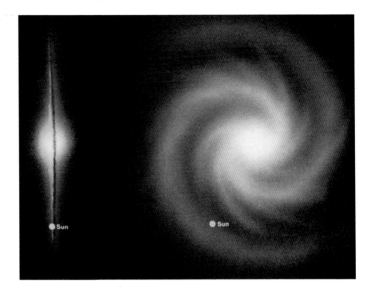

Our star-system or *Galaxy* contains about 100,000 million stars. It is a flattened system, and we lie well away from the centre; look along the main plane, and you will see many stars in almost the same direction, producing the lovely band of light which we call the Milky Way. Beyond our Galaxy, many millions of light-years away, we can see other galaxies, but I do not propose to say much about them here, because only one of them is easily visible with the naked eye from Britain: the Andromeda Spiral best seen on autumn evenings. Using powerful equipment, astronomers can now study systems so remote that we see them as they used to be long before the Earth was born.

THE PLANETS

The planets are our near neighbours – Venus, the nearest, can come within 39 million kilometres (24 million miles), so it is then no more than a hundred times as far away as the Moon – and this is why they wander around, though they keep to a well defined band known as the Zodiac. Luckily, they are easy to identify. Venus is so bright that it can cast shadows, and is visible either in the west after sunset or in the east before sunrise; at its best it looks almost like a small lamp. Because it is closer to the Sun than we are, it shows phases like those of the Moon, and these phases can be followed with good binoculars. Jupiter also is much brighter than any star. Mars is very red, and can be brilliant, though when furthest from us it does look like an ordinary red star. Saturn is bright (magnitude 1) and decidedly yellowish. If you are looking round the sky and come across a bright object which is not on your maps, you may be sure that it is a planet, but only Saturn and Mars can cause any real trouble; Venus and Jupiter give themselves away by their tremendous brightness.

◀ *Orion and surrounding stars appearing as sharp points of light. Betelgeux shows up clearly as orange-red.*

BUYING A TELESCOPE

Once you have started to find your way around, you may consider buying a telescope. A word of warning here! The thing not to do is to go out and buy a cheap telescope for a few tens of pounds. It may look attractive, but it is not likely to be of very much use astronomically.

Telescopes are of two types: *refractors* and *reflectors*. A refractor collects its light by means of a glass lens, while a reflector uses a specially curved mirror. The larger the main lens (for a refractor) or the mirror (for a reflector), the more powerful the telescope. It is the function of the lens or mirror to collect light, and the image is then magnified by a second lens or eyepiece. If you see a telescope advertisement in which only the 'magnification' is given, not the aperture, then ignore it.

I would advise against buying any refractor with a main lens less than 7.5 centimetres (3 inches) across, or any ordinary reflector with a mirror below 15 centimetres (6 inches) in aperture. Smaller telescopes do not collect much light, and are usually unsteady as well. To buy a telescope which will satisfy you means spending around £400 at least. It may sound a great deal – but compare it with the cost of a couple of train tickets between, say, London and Glasgow! Moreover, the cost is non-recurring; a telescope will last you a lifetime.

There are also 'catadioptric' telescopes, which are slightly more complicated, but are very good value. Meade and Celestron telescopes are particularly good.

The alternative is to buy a pair of binoculars. These are much cheaper, and £60 will provide a useful pair. Apart from sheer magnification, they have most of the advantages of a small telescope, and few of the drawbacks.

Armed with a pair of binoculars, you will be ready to start on a grand tour of the night sky. Even if you have no optical aid at all, and depend entirely upon your eyes, you will find plenty to interest you.

▲ *The Zodiac, passing through twelve constellations. (In fact a thirteenth, Ophiuchus, also crosses the Zodiac, between Scorpius and Sagittarius.)*

— *THE CONSTELLATIONS* —

The stars are divided up into constellations, some of which are very familiar while others are small and faint. The patterns we use are those of the Greeks, suitably modified. If we happened to use, say, the old Chinese or Egyptian patterns, our maps would look quite different – even though the stars themselves would be the same.

Because the stars are at very different distances from us, a 'constellation' has no real significance. For example, look at Ursa Major (the Great Bear), whose seven chief stars make up the pattern often nicknamed the Plough or the Big Dipper. The two 'end' stars are called Mizar and Alkaid (see page 22). They

WINTER

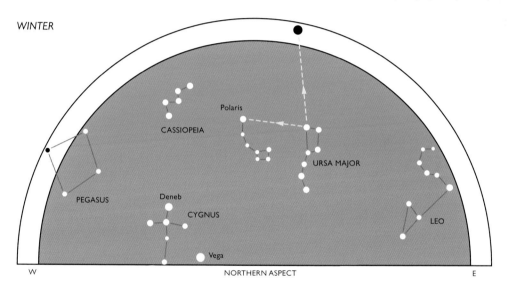

The winter sky in mid-evening

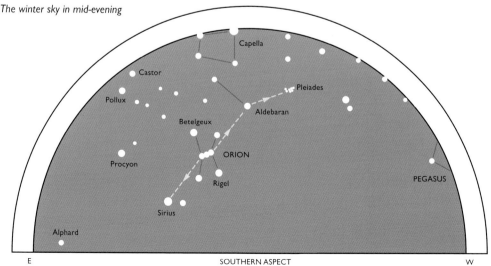

seem to be side by side, but they are not true neighbours; Alkaid is much further away than Mizar, so it merely lies in the same general direction as seen from Earth.

Following the Greek system, the constellations are named either after mythological gods and heroes (Hercules, Cepheus and Orion, for example), living creatures (such as the Swan or the Hare), or everyday objects (such as the Cup or the Triangle). Very few of the constellations have any resemblance to the objects after which they are named, and one great astronomer went so far as to comment that they seemed to have been designed so as to cause as much confusion and inconvenience as possible, but they will certainly not be altered now; we have become far too used to them.

SPRING

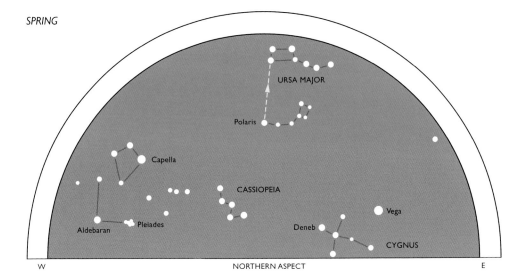

The spring sky in mid-evening

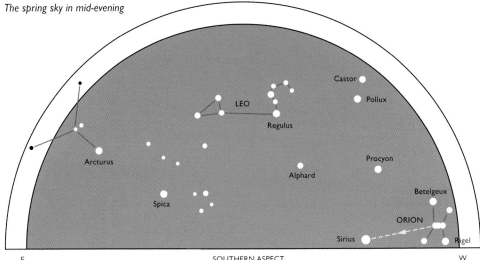

In general, the Latin names are used, since Latin is still the universal language; thus the Hare becomes 'Lepus' and the Swan 'Cygnus'. Individual stars have names of their own, usually of Arabic origin. These are used for the brightest stars – those of magnitude 1 – but not, in general, for fainter ones; astronomers have a different system, based upon the Greek alphabet, which need not concern us at the moment*.

The Earth takes one year to go round the Sun, so the Sun seems to move right round the sky in a period of one year, during which it passes through the twelve constellations of the Zodiac.

* Beware of an organization calling itself the 'International Star Registry', which claims to allot names to stars – naturally, upon payment of a substantial fee! This is quite valueless and unofficial, and is a complete waste of money.

SUMMER

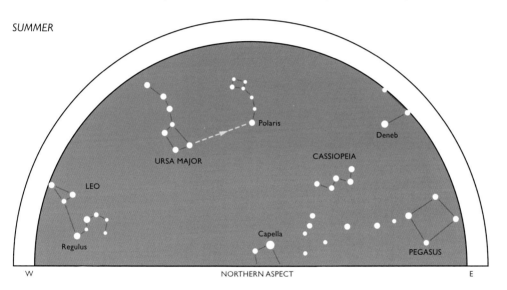

The summer sky in mid-evening

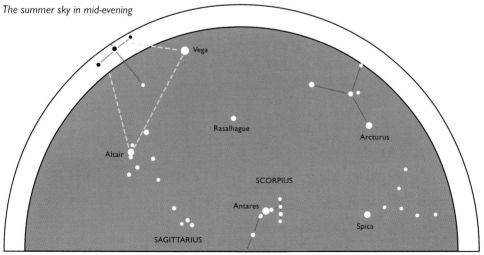

The Moon and bright planets are also found here, which is a great help in identifying them. They cannot wander into other regions; for example, you will never find a planet near the Great Bear.

I must pause briefly to dispose of astrology, the so-called 'science' which tries to link the positions of the planets with human character and destiny. There are still some people who take it seriously, but it has no basis whatsoever, and the only polite term to describe it is 'rubbish'. Never confuse astrology with astronomy.

NAVIGATING THE SKIES

My method of learning one's way around the night sky is to select a few groups which cannot be mistaken, and use these as guides to the rest. Our two best 'markers' are Ursa Major (the

AUTUMN

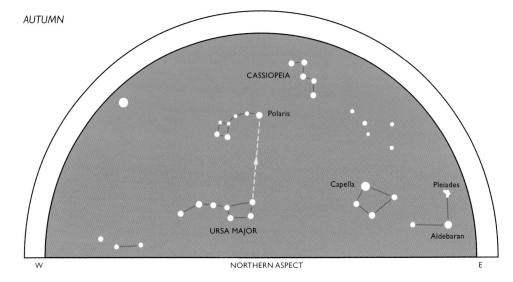

The autumn sky in mid-evening

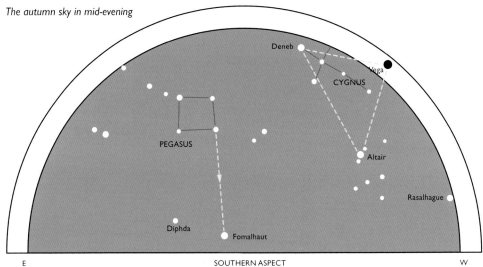

Great Bear) and Orion (the celestial Hunter). Ursa Major, with its Plough pattern, is particularly useful because it is circumpolar from Britain, and can always be seen somewhere or other whenever the sky is sufficiently clear and dark; Orion, crossed by the celestial equator, is not always above the horizon, and during summer it is too near the Sun in the sky to be seen at all.

What I have done over the last few pages, therefore, is to give seasonal charts – one for winter, one for spring, one for summer and one for autumn – in which I have put in the positions of a

THE MAIN CONSTELLATIONS VISIBLE FROM GREAT BRITAIN

Latin name	English name	Brightest star (magnitude)	Best seen during
Andromeda	Andromeda	Alpheratz (2.2)	Autumn
Aquarius †	The Water-bearer	—	Autumn
Aquila	The Eagle	Altair (0.8)	Summer
Aries †	The Ram	Hamal (2.2)	Autumn
Auriga	The Charioteer	Capella (0.8)	Winter
Boötes	The Herdsman	Arcturus (–0.04)	Spring
Camelopardus	The Giraffe	—	(circumpolar)
Cancer †	The Crab	—	Spring
Canes Venatici	The Hunting Dogs	Cor Caroli (2.9)	(almost circumpolar)
Canis Major	The Great Dog	Sirius (–1.5)	Winter
Canis Minor	The Little Dog	Procyon (0.4)	Winter
Capricornus †	The Sea-Goat	—	Autumn
Cassiopeia	Cassiopeia	—	(circumpolar)
Cepheus	Cepheus	—	(circumpolar)
Cetus	The Whale	Diphda (2.0)	Autumn/winter
Columba	The Dove	—	Winter
Coma Berenices	Berenice's Hair	—	(circumpolar)
Corona Borealis	The Northern Crown	Alphekka (2.2)	Spring
Corvus	The Crow	—	Spring
Crater	The Cup	—	Spring
Cygnus	The Swan	Deneb (1.2)	(mainly circumpolar)
Delphinus	The Dolphin	—	Summer
Draco	The Dragon	Eltanin (2.2)	(circumpolar)
Equuleus	The Foal	—	Summer
Eridanus	The River	—	Winter
Gemini †	The Twins	Pollux (1.1), Castor (1.6)	Winter
Hercules	Hercules	—	Summer
Hydra	The Watersnake	Alphard (2.0)	Spring
Lacerta	The Lizard	—	(circumpolar)
Leo†	The Lion	Regulus (1.4)	Spring
Leo Minor	The Little Lion	—	Spring
Lepus	The Hare	—	Winter
Libra †	The Balance	—	Spring
Lynx	The Lynx	—	(circumpolar)
Lyra	The Lyre	Vega (0.0)	Summer
Monoceros	The Unicorn	—	Winter
Ophiuchus	The Serpent-bearer	Rasalhague (2.1)	Summer
Orion	The Hunter	Rigel (0.2), Betelgeux (0.4)	Winter

Pegasus	The Flying Horse	—	Autumn
Perseus	Perseus	Mirphak (1.8), Algol (2.1)	Autumn/winter
Pisces †	The Fishes	—	Autumn
Piscis Australis	The Southern Fish	Fomalhaut (1.2)	Autumn
Sagitta	The Arrow	—	Summer
Sagittarius †	The Archer	Nunki (2.1)	Summer
Scorpius †	The Scorpion	Antares (1.1)	Summer
Scutum	The Shield	—	Summer
Serpens	The Serpent	—	Summer
Taurus †	The Bull	Aldebaran (0.8), Al Nath (1.6)	Winter
Triangulum	The Triangle	—	Autumn
Ursa Major	The Great Bear	Alioth (1.8), Alkaid (1.9), Dubhe (1.8), Mizar (2.3), Merak (2.3), Phad (2.4), Megrez (3.3)	(circumpolar)
Ursa Minor	The Little Bear	Polaris (2.0), Kochab (2.1)	(circumpolar)
Virgo †	The Virgin	Spica (1.0)	Spring
Vulpecula	The Fox	—	Summer

few key groups and stars. Using these, together with the more detailed maps in the main chapters, I hope that you will soon be able to become really familiar with the night sky.

† Denotes a zodiacal constellation.

THE BRIGHTEST STARS

Magnitudes are given to the nearest tenth; thus Altair (0.8) is brighter than Pollux (1.1). 'First-magnitude' stars are generally reckoned to be those down to magnitude 1.4. The two brightest stars visible from Britain, Sirius and Arcturus, have minus magnitudes; that of Arcturus is –0.04, which is nearer to zero than –0.1. Arcturus is therefore a full two magnitudes brighter than the Pole Star, which has a magnitude of 2.0.

THE BRIGHTEST STARS							
Star	Constellation	Magnitude	Colour	Star	Constellation	Magnitude	Colour
Sirius	Canis Major	–1.5	White	Bellatrix	Orion	1.6	White
Arcturus	Boötes	–0.0	Orange	Al Nath	Taurus	1.6	White
Vega	Lyra	0.0	Blue	Alnilam	Orion	1.7	White
Capella	Auriga	0.1	Yellowish	Alnitak	Orion	1.8	White
Rigel	Orion	0.1	White	Alioth	Ursa Major	1.8	White
Procyon	Canis Minor	0.4	White	Dubhe	Ursa Major	1.8	Orange
Betelgeux	Orion	0.4	Orange-red	Mirphak	Perseus	1.8	Yellowish
Altair	Aquila	0.8	White	Wezea	Canis Major	1.9	White
Aldebaran	Taurus	0.8	Orange	Alkaid	Ursa Major	1.9	White
Antares	Scorpius	1.0	Red	Menkarlina	Auriga	1.9	White
Spica	Virgo	1.0	White	Alhena	Gemini	1.9	White
Pollux	Gemini	1.1	Orange	Alphard	Hydra	2.0	Orange
Fomalhaut	Piscis Australis	1.2	White	Mirzam	Canis Major	2.0	White
Deneb	Cygnus	1.2	White	Algieba	Leo	2.0	Orange
Regulus	Leo	1.3	White	Polaris	Ursa Minor	2.0	White
Adhara	Canis Major	1.5	White	Hamal	Aries	2.0	Orange
Castor	Gemini	1.6	White				

THE WINTER SKY

Winter is a good time to begin a study of the sky. This is because both the most helpful 'guides', Ursa Major and Orion, are on view during winter evenings – Ursa Major in the north-east, Orion high in the south.

URSA MAJOR, URSA MINOR AND CASSIOPEIA

Ursa Major looks nothing like the shape of a bear, though I suppose that with a considerable effort of the imagination it is possible to make a plough or a dipper out of it! Six of the stars (Alioth, Mizar, Alkaid, Dubhe, Merak and Phad) are around second magnitude, while the seventh, Megrez, is a magnitude fainter.

▼ *Artist's impression of Mizar and Alcor as seen through a Telescope. Mizar is itself double; Alcor is to the lower right. The star to the upper right is 'in the background', and is not associated with the Mizar-Alcor group.*

Two of the Plough stars are worthy of special note. First, look at Mizar. Close beside it you should be able to make out a much fainter star, Alcor. They make up a genuine pair, even though they are a long way apart; an old nickname for them is 'Jack and his Rider'. Strangely enough, the Arab stargazers of a thousand years ago regarded Alcor as a test of keen eyesight, but there is nothing difficult about it today if the sky is reasonably clear and dark. Binoculars show the pair well, and if you

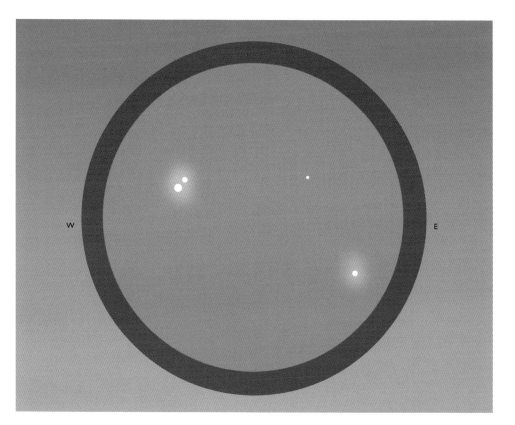

W E

have a telescope you will see that Mizar itself is made up of two stars, which are so close together that with the naked eye they appear as one.

Double stars are common in the sky. Sometimes they are due to a chance lining-up, but in most cases the two members of the pair are genuinely associated, and are moving round their common centre of gravity much as the two bells of a dumbbell will do if you twist them by the bar joining them. Mizar is a 'binary system' of this type, but the two components are so widely separated that the revolution period amounts to many thousands of years.

Merak and Dubhe are nicknamed the Pointers, because they show the way to the Pole Star. Dubhe is decidedly orange, showing that its surface is relatively cool – even though it is a large star around 60 times as luminous as the Sun. The orange hue can be noticed with the naked eye, but binoculars bring it out much more strongly. If you look first at Dubhe and then at Merak, you will see what I mean.

To find Polaris, take a line from Merak, pass it through Dubhe and continue until you come to Polaris, which is exactly magnitude 2 (slightly fainter than Dubhe, slightly brighter than Merak). As we have found, Polaris seems to stay almost still, with everything else rotating round it once in 24 hours. It is 680 light-years away, so we are now seeing it as it used to be in the reign of Edward II.

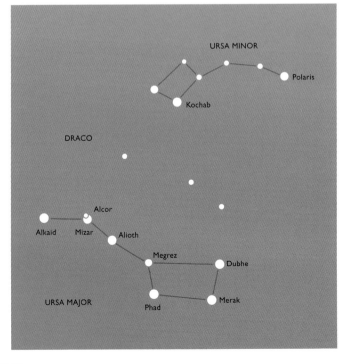

◀ The Plough and the Pole Star, Polaris.

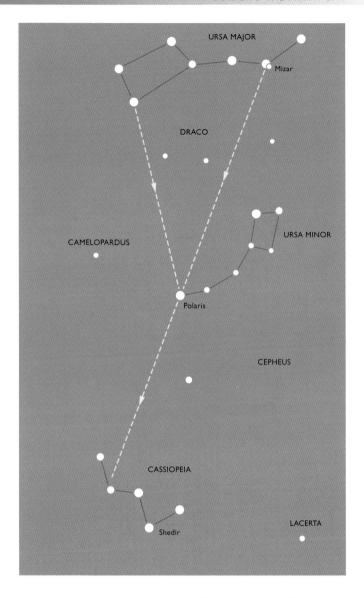

▶ *Finding Cassiopeia by using Mizar and Polaris as 'guides'.*

Next, look from Polaris in the direction of Alkaid, the end star in the Great Bear's 'tail'. You will see a pattern which is a little like that of the Great Bear, but is much fainter and rather distorted. This is *Ursa Minor*, the Little Bear. It contains one more second-magnitude star, Kochab, but the rest are fainter, and moonlight or haze will hide them. Binoculars show that Kochab, like Dubhe, is decidedly orange in colour.

Follow on the line from the Pointers through Polaris, and you will arrive somewhere near a W or M formation of stars, of which the brightest are of magnitude 2. This is *Cassiopeia*. Four of the stars in the W pattern are white; the fifth, Shedir, is orange.

Cassiopeia is a rich constellation, because the Milky Way flows through it. Binoculars show endless star-fields, and it is easy to believe that the stars in the Milky Way are in grave danger of colliding with each other. Of course this is not so; we are dealing with another line-of-sight effect, and the average distance between stars amounts to several light-years.

It is worth remembering that the Great Bear, Polaris and Cassiopeia are approximately lined up, so when the Bear is high up Cassiopeia is low down – and vice versa. All these groups are circumpolar from Britain. So too are some other constellations of lesser note, such as *Draco* (the Dragon), *Camelopardus* (the Giraffe), *Lynx* (the Lynx), *Lacerta* (the Lizard), *Cepheus, Canes Venatici* (the Hunting Dogs), *Coma Berenices* (Berenice's Hair) and most of *Leo Minor* (the Little Lion).

ORION AND SURROUNDINGS

Now it is time to look for our second main guide, *Orion*, which dominates the evening sky throughout the winter. In mythology, Orion was a great hunter. In the sky he stands out, partly because of his characteristic shape and partly because his stars are so bright.

Two of the stars in Orion are particularly brilliant. The first is Betelgeux, in the Hunter's shoulder – the upper left star of the main quadrilateral. (The name can be pronounced in various ways. Some people call it 'Beetlejuice'.) It is first magnitude, and is orange-red, as any casual glance will show. This means that it is cooler than our yellow Sun; but to make up for this, Betelgeux is huge. Its diameter is 400 million kilometres (250 million miles) so it could swallow up the whole path of the Earth round the Sun. Yet it is over 500 light-years away, and looks like nothing more than a speck of light.

Betelgeux is an unstable star. It swells and shrinks, and as it does so its magnitude changes. It is always bright, and at its best it may become almost the equal of Rigel, though at minimum it is little brighter than Aldebaran in the Bull.

Rigel, in the Hunter's foot, is different. It is pure white, and is exceptionally luminous; take 60,000 Suns, put them together, and you have one object as powerful as Rigel.

The three bright stars lined up along the centre of the main pattern make up the Hunter's Belt; Alnilam and Alnitak are just above second magnitude, Mintaka just below. Just below the Belt you should be able to make out a misty patch. This is known as a nebula, from the Latin word meaning 'cloud'. It is made up of dust and gas, and is a stellar nursery; inside it, fresh stars are being produced – though the process is a slow one. Around 5000 million years ago, our Sun was born inside a nebula of this kind.

Nebulae are common enough, but not many are visible with the naked eye, and the Orion Nebula in the Hunter's Sword is the most famous example. Binoculars show it well, and with a telescope you can see the wispy gas with darker patches here and there. It is well over a thousand light-years away.

Orion is crossed by the Milky Way, and the whole area is very rich. The other two stars in the main pattern are Bellatrix (upper right) and Saiph (lower left), both second magnitude and both hot and white.

▼ *Orion and surroundings.*

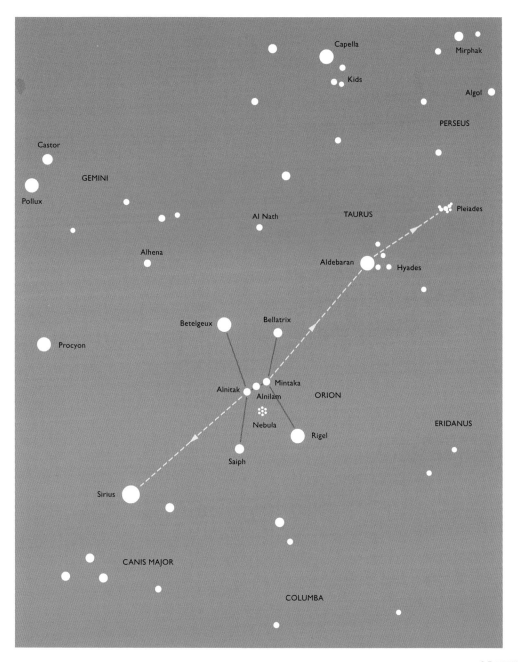

Follow the line of the Belt downward, and you will come to Sirius, which is by far the brightest star in the sky (though it cannot rival the most brilliant planets, Venus or Jupiter). Sirius is the leader of *Canis Major*, the Great Dog. It looks much more striking than Rigel or Betelgeux, but this is only because it is so much nearer to us. It is in fact one of the closest stars in the sky; it is a mere 8.6 light-years away, and is only 26 times as powerful as the Sun.

Sirius is white, but it seems to flash various colours of the rainbow, and in binoculars or a telescope it is a glorious sight. This is due entirely to the fact that its light is coming to us through the Earth's unsteady atmosphere. Twinkling has nothing directly to do with the stars themselves, but it is very noticeable, and Sirius is the supreme 'twinkler', partly because it is so bright and partly because as seen from Britain it is always rather low down. The higher up a star is in the sky, the less it twinkles.

There are several other bright stars in Canis Major, and it too is crossed by the Milky Way. Higher up you will see Orion's second Dog, *Canis Minor*, which has one first-magnitude star, Procyon. Like Sirius, Procyon is white and relatively close to us; its distance is just over 11 light-years, which works out to something like 97 million million kilometres (60 million million miles).

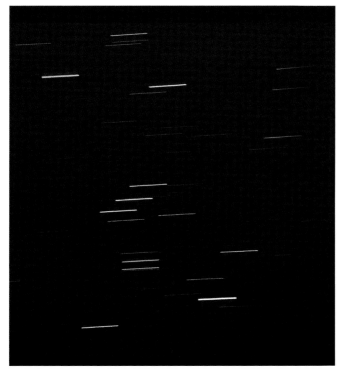

◄ *Orion. This is a time-exposure, and the camera was not being rotated, so the stars show up as trails. The orange-red trail to the upper left is Betelgeux. The Great Nebula, below the belt, also shows up as orange, though with the naked eye or in binoculars it looks white.*

► *The Pleiades, or Seven Sisters — the most famous star cluster in the sky.*

Follow the line of Orion's Belt upward, and you will come to a bright orange-red star. This is Aldebaran, leader of *Taurus*, the Bull. Extending to its right is a little ∨-formation of stars; these make up a loose cluster, the Hyades. Oddly enough, Aldebaran is not a true member of the cluster. It simply happens to lie in much the same direction, and is about midway between the Hyades and ourselves. If we happened to be observing from another vantage point in the Galaxy, Aldebaran and the Hyades could be seen on opposite sides of the sky.

Aldebaran is about the same colour as Betelgeux, but it is much less luminous. It is interesting to compare the two, though Betelgeux is generally rather the brighter.

Keep on with the line from the Belt through Aldebaran; continue upward, and then curve down a little. You will see a small, hazy patch. This is the cluster of the *Pleiades* or Seven Sisters. Look more carefully, and you will realize that it is made up of stars. On a clear night you should be able to make out at least seven stars, and keen-eyed people can see more; the record is said to be nineteen. Binoculars show dozens, and the total number of stars in the cluster is about four hundred.

Thousands of these loose clusters are known, but the Pleiades is the brightest of them all. Most of its leading members are hot and bluish-white, which means that by stellar standards they are young.

Almost at the zenith or overhead point you will see a very bright star, Capella in *Auriga*, the Charioteer. It is yellow, like the Sun, but much more luminous. Auriga contains one other bright star, Menkarlina (magnitude 2), and is easy to identify, because the constellation has the shape of a rather distorted quadrilateral; note too the triangle of faint stars closely right of Capella, often nicknamed 'the Kids'. Almost directly between Capella and Bellatrix, in Orion, is yet another second-magnitude star, Alnath. It used to be included in Auriga, but for some reason or other the International Astronomical Union, the controlling body of world astronomy, gave it a free transfer into Taurus!

Orion's main retinue is completed by the twins, Castor and Pollux in *Gemini*, which lie side by side. A line from Rigel through Betelgeux, continued for some way, will reach them. Pollux is rather the brighter of the two, and is orange, whereas Castor is white. The rest of Gemini consists of lines of stars extending from Castor and Pollux in the general direction of Orion; one of them, Alhena, is second magnitude.

Sirius, Procyon, the Twins, Capella and Aldebaran make up a huge curve around Orion. The Milky Way runs across the whole area, and is very rich in the large but faint constellation of *Monoceros*, the Unicorn. Beyond Auriga, the Milky Way runs on to *Perseus*, with the second-magnitude Mirphak, and thence Cassiopeia.

Finally, look for some of the other constellations shown on the winter map: the rather blank *Eridanus* (the River) in the south-west, and *Lepus* (the Hare) below Orion, with the Square of Pegasus almost setting, and *Leo* (the Lion) coming into view in the east. Pegasus is best seen during the autumn evenings and Leo in the spring, so I will have more to say about them later.

THE SPRING SKY

By spring we are looking forward to the summer, and to what we hope are going to be long days and warm nights, but in spring there is still a good deal of darkness. The aspect of the evening sky has changed since winter. We are losing Orion, though much of it is still visible low in the west; Leo, the Lion, is high in the south, while Capella is descending in the west and Vega rising in the east. Ursa Major, the Great Bear, is almost overhead, and this means that the W of Cassiopeia is at its lowest, though it is still well above the horizon.

USING URSA MAJOR AS A GUIDE

Let us start our spring survey with Ursa Major. (I know that this means tipping yourself backwards – so why not fetch a comfortable chair and sit down?) As before, follow the line of the Pointers and locate the Pole Star. Next, follow round the curve of the Bear's 'tail'. Before long you will come to a very brilliant star, Arcturus in *Boötes* (the Herdsman). It is impossible to overlook Arcturus, partly because it is so bright and partly because it is of a lovely orange hue, brought out even more

▼ *Spring group: using Ursa Major as a guide.*

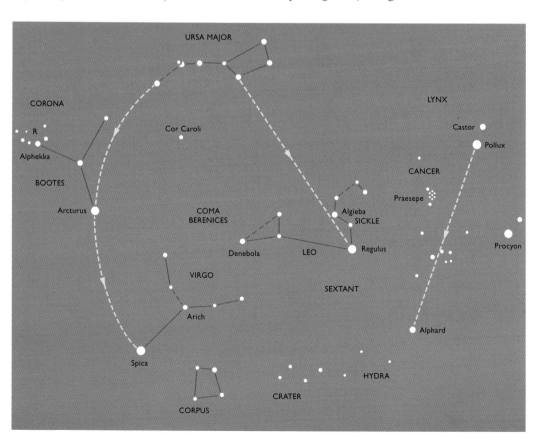

strongly with binoculars. Arcturus is 115 times as luminous as the Sun, and is 36 light-years away – much closer than Polaris or Rigel, but more than four times as far as Sirius.

It is not easy to make a human figure out of the pattern of Boötes, but you will find a kind of large Y formation, with Arcturus at the bottom of the 'stem'. The upper left of the Y is not in Boötes, but in the adjacent constellation of *Corona Borealis*, the Northern Crown, whose leader is the second-magnitude Alphekka.

Corona is made up of a little semicirclet of stars. Look at it with binoculars, and you should see that inside the 'bowl' of the Crown there are two stars. The upper one is a very ordinary star of magnitude 6.5, so it is just too faint to be seen with the naked eye. The other – toward the lower part of the inside of the bowl – is a very strange star, known by its official title of R Coronae. Usually it is easy to see with binoculars, but sometimes clouds of soot build up in its atmosphere, and it hides itself behind a thick veil, so you need a large telescope to see it at all. It may stay dim for weeks before the soot blows away. If you can see only one star in the bowl, you may be sure that R Coronae has faded; look again later, and it will probably be back. Variable stars of this kind are unusual, and R Coronae is the brightest member of its class. We may be thankful that our placid Sun does not behave in such a way.

If you follow the curve of the Bear's tail through Arcturus, and then continue it, you will come to another first-magnitude star, low over the south-east horizon during spring evenings. This is Spica, in *Virgo*, the Virgin. It looks much fainter than Arcturus, though it is actually more powerful, and is the equal of over 2000 Suns; it is 210 light-years away.

Virgo is a very large constellation. It too is shaped rather like a Y, but this Y has a definite bowl, not too unlike the pattern of Boötes. The star at the base of the bowl, above and to the right of Spica, is named Arich, which is third magnitude. Telescopes show that it is a binary, with the two members exactly alike – a pair of stellar twins. On the far side of the bowl, well away from Spica and to the right, is the second-magnitude Denebola, in Leo.

The area enclosed by Virgo, Denebola, Arcturus and Ursa Major contains no bright stars, apart from Cor Caroli in *Canes Venatici* (the Hunting Dogs). The region looks like a large, faint star-cluster. This is the region of *Coma Berenices*, or Berenice's Hair, and is worth sweeping with binoculars.

Before passing on, look below Spica and rather to the right. Not far above the horizon it is easy to find a quadrilateral of stars, all around third magnitude. These make up *Corvus*, the Crow. It contains little of real interest, but it is not hard to locate, because there are no other even reasonably bright stars anywhere near it. To its right is another small constellation, *Crater* (the Cup), which has no star much brighter than fourth magnitude.

LEO, HYDRA AND CANCER

The main spring constellation is *Leo*, the Lion. To find it, come back to Ursa Major, which remember is almost straight overhead. Use the Pointers the 'wrong way', away from Polaris, to find Leo. More accurately, use the other two stars in the quadrilateral of the Bear, Megrez (the faint one) and Phad (lower left). Direct a line downward, and you will come to a curved line of stars, of which the brightest, Regulus, is first magnitude. This is Leo.

The Lion is large. The curved line, nicknamed the Sickle, is shaped rather like the mirror image of a question-mark. The second star above Regulus is called Algieba; it is orange, though the colour is not so obvious as with Arcturus because Algieba is so much fainter. Telescopes show that it has a fainter, rather bluish companion.

Over to the left of the Sickle, at about the same height above the horizon, is a triangle of stars making up the rest of the Lion. The brightest of the triangle is Denebola, which, as we have seen, is on the far (right) side of the bowl of Virgo. Denebola is rather interesting. Astronomers of a couple of thousand years ago made it as bright as Regulus, but it is now a full magnitude fainter. Either it has faded, or else more probably the old stargazers were wrong.

Between Leo and the southern horizon there is a rather blank area. This is the region of *Hydra*, the Watersnake, which is the largest constellation in the entire sky, but contains only one brightish star. This is the second-magnitude Alphard, known as 'the Solitary One' because it is so much on its own. One way to locate it is to take a line from Alkaid, in the Great Bear, pass the line through Regulus and continue until you reach Alphard, which is decidedly orange. But there is another method and this brings us back to the Orion area.

During spring evenings the top part of Orion is still on view; you should be able to find Betelgeux, still well above the horizon, and probably the three stars of the Belt, though Rigel has almost set. Of the Hunter's retinue, you can still just about see Sirius; above and to the left of Sirius is Procyon in Canis Minor, and there should be no problem in finding the Twins, Castor and Pollux. Take a line from Castor, pass it through Pollux, and continue for some way. The first bright star you will reach is Alphard.

(To check on the Twins, it is also useful to go back to the Great Bear. A line from Megrez through Merak, the fainter of the two Pointers, will lead you to Castor and Pollux.)

Between the Twins and Alphard – and also between Procyon and Regulus – is the faint Zodiacal constellation of *Cancer*, the Crab. Cancer has no bright stars, but you will be able to make out a little misty patch, almost in the middle of the large triangle formed by Regulus, Pollux and Procyon. This is the loose star-cluster Praesepe, often nicknamed the Beehive. Binoculars show it well, though it is much less rich than the Pleiades. To either side of Praesepe you will see two dim stars, known as the 'Asses' because another nickname for Praesepe is 'the Manger'.

Remember that Cancer, Leo and Virgo are all in the Zodiac. If you see a bright object there which is not shown on the maps, you may be sure that it is a planet – either Mars, Jupiter or Saturn.

Capella, which was so near the zenith during winter evenings, is now in the north-west; between it and Ursa Major is the faint but large constellation of *Lynx*. Over in the north, look for the W of Cassiopeia; one good way to find it is to take a line from Mizar, in the Great Bear, through Polaris – continue this line, and you will come to Cassiopeia. Rising in the north-east is another brilliant star, the bluish Vega, which will be almost overhead during summer evenings. It is worth noting that Vega and Capella lie on opposite sides of the Pole Star, and at about the same distance from it. If you see a bright star near the zenith, it can only be one of these two.

THE SUMMER SKY

The short summer nights may not be ideal for stargazing, but at least they are warm, and there is no need to wrap up. During summer evenings Ursa Major is in the north-west; from it, find Polaris and, in the north-east, the W of Cassiopeia. Follow round the 'tail' of the Bear and you will come to Arcturus, the brightest star in the northern hemisphere of the sky (its only three superiors, Sirius, Canopus and Alpha Centauri, are all in the southern hemisphere, and neither Canopus nor Alpha Centauri can ever be seen from Britain). The Υ of Boötes, together with Alphekka in the Northern Crown, is very obvious, but Spica and the rest of Virgo are setting, and to all intents and purposes we have lost Leo. Of Orion there is no sign at all; it is much too near the Sun in the sky, and is above the horizon only during daylight.

VEGA, DENEB AND ALTAIR

Almost overhead lies Vega, the leader of Lyra, the *Lyre* or Harp. It is a lovely star; to me it always looks decidedly blue. It is 50 times as luminous as the Sun, and is 26 light-years away. It is of special interest because we have found that it is surrounded by clouds of cool material which might indicate a system of planets – though this material cannot be seen directly, and has to be studied by special techniques.

Lyra is a small constellation, but it contains several objects of real interest. Close beside Vega is Epsilon Lyrae, which is a double star; keen-eyed people can separate the two components without optical aid, and binoculars show them well. Telescopically, it is found that each component is again double, so Epsilon Lyrae is a quadruple system. Delta Lyrae, also near Vega, is a double of a different type; the brighter member of the pair is red, and the fainter member white. With any pair of binoculars Delta Lyrae is striking, particularly as it lies in a very rich field.

Look next at two more of the stars near Vega, Sheliak and Sulaphat. Sulaphat is just below third magnitude; Sheliak is variable, with a range of from just below magnitude 3 to just below magnitude 4. In fact Sheliak is made up of two stars, too close to be seen separately, moving round their common centre of gravity. The apparent magnitude changes as the two components pass behind and in front of each other.

Directly between Sheliak and Sulaphat is the famous 'Ring Nebula', which is not a true nebula, but an old star which has thrown off its outer layers. It is too faint to be seen with the naked eye, and I have never been able to find it with binoculars, but telescopes show it easily; it looks rather like a tiny, dimly shining cycle tyre.

Over to the east of Vega, also very high up, is Deneb in *Cygnus*, the Swan. It is not nearly so bright as Vega, but this is only because it is much further away; it has at least 70,000 times the luminosity of the Sun, and we see it now as it used to be at the time of the Roman Occupation. Lower down, in the south-east, is yet another first-magnitude star, brighter than Deneb though not nearly so brilliant as Vega. This is Altair, leader of *Aquila*, the Eagle. Altair is particularly easy to recognize because it has a fainter star to either side of it; Tarazed above, Alshain below.

Vega, Deneb and Altair form a huge triangle. Many years ago, in a television *Sky at Night* programme, I nicknamed this 'the Summer Triangle', and the term has come into general use, though it is completely unofficial – and does not apply in southern countries such as Australia, where it is best seen in winter.

Both Cygnus and Aquila are crossed by the Milky Way, so they are very rich. Cygnus is often nicknamed the Northern Cross, for obvious reasons. From Deneb, extending to the area between Vega and Altair, there is a striking ✕-pattern. It is not symmetrical; the central star, Sadr, is second magnitude, but one of the stars in the ✕, Albireo, is fainter than the rest and is

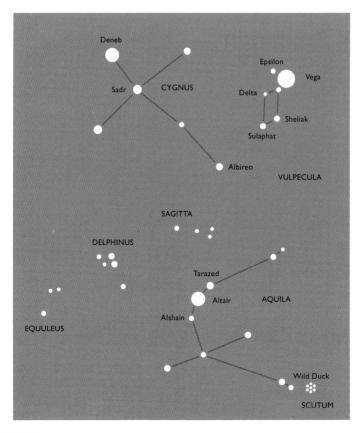

◄ *Vega, Deneb and Altair: the 'Summer Triangle'.*

also further away from the centre of the cross. Join Vega and Altair with a line; find the centre point, and slightly 'up and to the right' you will see Albireo.

With the naked eye, Albireo looks like an ordinary star. With powerful binoculars, or a telescope, it is seen to be double; the bright member of the pair is golden yellow, while the companion is vivid blue. In my view, this is the most beautiful coloured double star in the whole of the sky.

Aquila is almost as rich, and the pattern is easy to identify; it is not hard to picture a flying bird. There are various small constellations in this area, notably *Delphinus*, the Dolphin, which is small and compact (unwary observers have even mistaken it for the Pleiades). *Sagitta*, the Arrow, is also easy to find, though its stars are dim. *Vulpecula*, the Fox, and *Equuleus*, the Foal, are much less obvious. Vulpecula used to be known as Vulpecula et Anser, the Fox and Goose, but on modern maps the goose has disappeared.

At the lower end of Aquila is another small constellation, *Scutum*, the Shield. It has no bright stars, but it does contain a bright star-cluster, which is nicknamed the Wild Duck. It is somewhat fan-shaped, and binoculars show it well, though it is not always easy to pick out because the Milky Way near it is so very rich.

SCORPIUS AND SAGITTARIUS

Very low over the southern horizon there are two Zodiacal constellations, *Scorpius* (the Scorpion) and *Sagittarius* (the Archer). Both are superb, but unfortunately they are never seen to advantage from Britain; parts of them remain below the horizon, and from North Scotland they are not easy to see at all. The brightest star in Scorpius is the red first-magnitude Antares. Like Altair, it is flanked to either side by a fainter star, but there is no danger of confusing the two, because Antares is so strongly coloured and is also much lower down.

It is worth noting that Arcturus, Vega and Antares form yet another huge triangle. Moreover, the Milky Way flows from Cygnus and Aquila through Scutum and then into Scorpius and Sagittarius.

Not many constellations bear much resemblance to the objects they are supposed to represent, but Scorpius is an exception; it has a 'head', and then a long line of stars in which Antares is outstanding. Antares is the reddest of all the bright stars, and, like Betelgeux in Orion, is large enough to contain the whole path of the Earth round the Sun. It is 330 light-years away, and is the equal of 7500 Suns.

Following Scorpius round, even lower over the horizon, is Sagittarius. It has a number of brightish stars, notably the second-magnitude Nunki, but there is no really distinctive pattern, though it is often likened to a teapot. The Milky Way is at its very richest in Sagittarius, and the star-clouds are easily visible with the naked eye as luminous patches. It is well worth sweeping the area with binoculars.

As we have seen, our Galaxy is a flattened system, and the Sun, together with the Earth and the other planets, is about 30,000 light-years from the centre. The heart of the Galaxy is never visible, because there is too much interstellar 'dust' in the way, but we know that it lies beyond the star-clouds of the Archer. The Sun is moving round the centre of the Galaxy, taking about 225,000,000 years to make one circuit – a period often termed the 'cosmic year'. One cosmic year ago, the most advanced life-forms on Earth were amphibians; even the dinosaurs had yet to make their entry. It is interesting to speculate as to what the world will be like one cosmic year hence.

▼ *The Hercules region, with Scorpius and Sagittarius.*

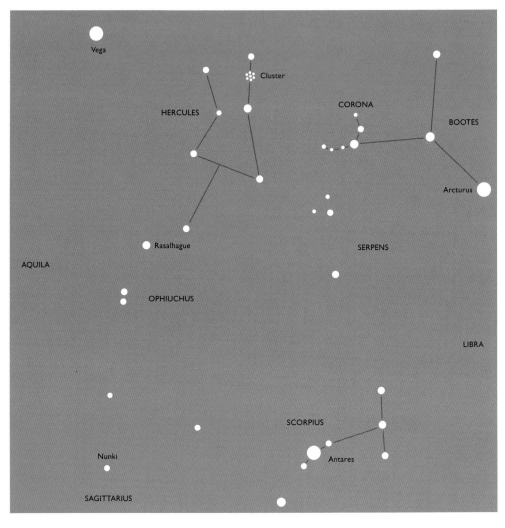

▶ *Globular cluster.*

THE HERCULES REGION

Next, come back to Vega, almost overhead. Relocate Arcturus, by following round the 'tail' of the Great Bear. We have just found Antares, low in the south-west. Inside the triangle outlined by these three bright stars is a rather dull area, occupied by three large constellations – *Hercules*, the legendary hero; *Ophiuchus*, the Serpent-bearer, and *Serpens*, the Serpent, with which Ophiuchus is struggling. Serpens appears to be having the worst of the encounter, and is split into two separate parts, a Head and a Body.

Frankly, these three groups are not too easy to sort out. The only star in them as bright as second magnitude is Rasalhague, the leader of Ophiuchus, which lies not far off a line joining Vega to Antares. However, there is one object of special interest in Hercules: a cluster of quite different type from the Pleiades or the Wild Duck. It is a symmetrical system – a 'globe' of stars, which is why it is termed a globular cluster. It may contain as many as a million suns, but it is 22,000 light-years away, so is on the fringe of naked-eye visibility. Binoculars show it easily, and a telescope shows that it really is made up of stars.

About a hundred globular clusters are known, but only three are naked-eye objects, and the other two are in the far south, so from Britain they never rise. The globulars lie round the edges of the main Galaxy, and seem to be very old by stellar standards, so their leading stars are reddish. Near the centre of a globular cluster the stars are much closer together than in other regions of the Galaxy. If our Sun lay in such a cluster, there would be many stars bright enough to cast shadows, and there would be no real darkness at night.

Over in the east the Square of Pegasus is rising, and by the early hours of the morning it is reasonably high, but it is mainly an autumn group. Finally, low in the south-west, is *Libra*, the Balance or Scales, which is one of the faintest of the Zodiacal constellations, and contains nothing of immediate interest.

METEORS AND COMETS

Summer is the best time for seeing shooting-stars or meteors. A meteor is a tiny particle, usually smaller than a pin's head, moving round the Sun. If it dashes into the Earth's upper air, moving at a speed of perhaps 72 kilometres (45 miles) per second, it has to force its way through the air particles. This sets up friction; friction causes heat, and the particle burns away in the streak of radiance which we call a shooting-star.

Between 27 July and 17 August the Earth passes through a swarm of these particles, and the result is a shower of shooting-stars. They seem to come from the direction of Perseus, which is now low in the north, and they are therefore known as the Perseids. Meteors may be seen at any time, and there are a number of annual showers, but the Perseids are particularly rich, and can usually be relied upon to give a good dis-

▶ *Brilliant meteor. A meteor plunges downward, burning itself away in the atmosphere.*

▼ *Meteor radiant. The meteors from any particular shower seem to come from one special point or 'radiant' in the sky; the radiant of the August shower is in Perseus. Of course, the meteors are not seen at the same time, but if their paths are plotted 'backwards' they meet at the radiant.*

play. Look up into a dark, clear sky at any time during this period, and you will be unlucky not to see a meteor or two. The shower reaches its peak on 12 August. Obviously, strong moonlight will interfere, so some displays are more spectacular than others.

Meteors are associated with comets, which are the most erratic members of the Solar System. A comet is made up of an icy centre or nucleus, only a few kilometres or miles across, surrounded by a 'head' of small particles and very tenuous gas. As it moves, it leaves a dusty trail behind it, and it is this trail which produces meteors. The comet associated with the Perseids is known as Swift–Tuttle, after its discoverers. It takes well over a hundred years to go round the Sun, and will not next be reasonably close to us until the year 2127.

Because a comet is millions of kilometres or miles away, it does not shift quickly against the background of stars, and has to be watched for hours before any motion can be detected. If you see an object travelling perceptibly across the sky, it cannot be a comet. The most famous of all comets, and the only bright one which we can predict, is named in honour of Edmond Halley, Britain's second Astronomer Royal, who died as long ago as 1742. Halley's Comet returns every 76 years; it was last close to us in 1986, though unfortunately it was badly placed, and never became as bright as it had done at the previous returns of 1910 and 1835.

Now and then the Earth collides with a larger object, which can survive the complete drop to the ground without being burned away, and is then called a meteorite. However, meteorites are not simply large meteors, and they are not cometary debris; they come from the belt of asteroids or minor planets between the orbits of Mars and Jupiter, or from Mars. On rare occasions the Earth may be hit by a meteorite large enough to produce a crater, but there is no reliable record of any human death due to a meteorite impact – and I can assure you that if you watch the Perseids on a summer evening, you are in no danger of being hit on the head by a falling stone!

—— *THE AUTUMN SKY* ——

It has often been said that the autumn skies are less spectacular than those of other seasons. In a way this is true, if only because Orion does not rise until shortly before sunrise, and Ursa Major is at its lowest in the north; Arcturus has set, and so have Scorpius and Sagittarius. However, we still have the 'Summer Triangle'. Vega is in the north-west; Deneb is higher up; and Altair is well above the horizon. The W of Cassiopeia is almost overhead, with Capella rising in the north-east.

Then, too, we have the Pleiades. The Pointers in Ursa Major together with Capella show the way to the 'Seven Sisters'; I always feel that the sight of the Pleiades rising eastward in the evening sky is a sure sign that summer is over, and that winter, with its frosts and fogs, lies ahead. Lower still, very near the horizon, you can see Aldebaran.

CASSIOPEIA AND PERSEUS

To locate the main autumn groups, it is again convenient to start with Ursa Major. A line from Mizar through Polaris, and continued for about the same distance on the far side, brings us to *Cassiopeia*. Between Cassiopeia and Capella is *Perseus*, shaped rather like an inverted and distorted Y; its leader,

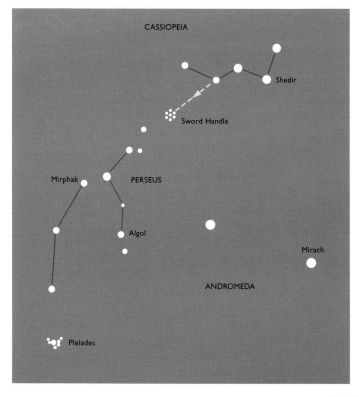

▶ *Cassiopeia and Perseus.*

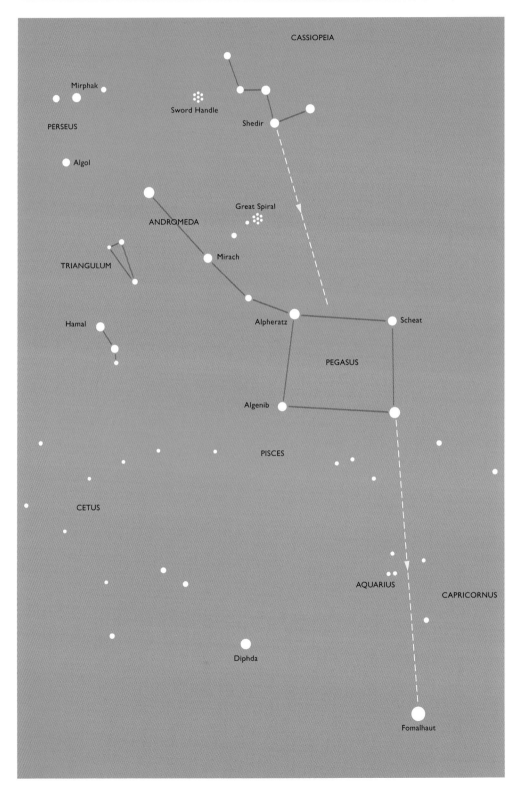

Mirphak

◄ Pegasus and Fomalhaut.

Mirphak, is second magnitude. The rest of Perseus stretches down in the general direction of the Pleiades.

There are two objects here of special interest. One is Algol, the second star of the constellation. Usually it is second magnitude, but every 2.8 days it fades down to below magnitude 3. It takes four hours to fall to minimum, after which it stays faint for a mere twenty minutes and then takes another four hours to recover. It is the brightest of the 'eclipsing binaries', and is not genuinely variable; minimum is caused when the fainter of the two Algols passes in front of the brighter and cuts out part of its light.

There is a famous legend associated with Perseus. It is said that he was a dashing hero who had cut off the head of the Gorgon, Medusa, who had a woman's face but hair of snakes, and whose glance would turn any living creature to stone. On his return, Perseus came across the Princess Andromeda, who had been chained to a rock on the sea-shore as a sacrifice to a fearsome monster. Perseus turned the monster to stone by showing it the Gorgon's head, and then, in the best tradition, married Andromeda and 'lived happily ever after'. In the sky, Medusa's head is marked by Algol, which has long been known as the Demon Star. Whether the astronomers of ancient times knew about Algol's strange 'winks' is uncertain, but they may well have done.

The other special feature of Perseus is a double star-cluster, known as the Sword-Handle. To find it, use two stars of Cassiopeia as guides, as shown in the diagram on page 41. The Sword-Handle is visible with the naked eye as a dim, misty blur; binoculars show that there are two separate star-clusters side by side, and in any telescope they are truly magnificent.

PEGASUS AND FOMALHAUT

The main autumn constellation, *Pegasus* (the Flying Horse), is now high in the south. It is very distinctive, and its four chief stars are arranged in the pattern of a square, but it is not particularly bright, and maps tend to make it look smaller and more prominent than it really is. One way to identify it is to use Cassiopeia as a guide. Take a line from the centre star of the W, pass it through Shedir (the orange member of the group) and continue it; it will lead straight to the Square.

The second-magnitude upper left-hand star of the Square has a proper name – Alpheratz. It joins Pegasus on to the adjacent constellation, *Andromeda*, and is officially included in that constellation, which seems illogical because it so clearly fits in with the Pegasus pattern. Andromeda is marked by a line of four stars, of which Alpheratz is the first, leading along to Mirphak in Perseus, so Alpheratz, the Andromeda line, Mirphak and Capella are almost lined up.

(Incidentally, see how many naked-eye stars you can see inside the Square on a really dark night. You may be surprised.)

▲ *The Andromeda Galaxy.*

The most famous object in Andromeda is the Great Spiral, known officially as Messier 31. It is a separate galaxy, and is the most remote object clearly visible without optical aid. To find it, identify the stars in the Andromeda line; Alpheratz is the first, the second is rather fainter (magnitude 3.5) but the third, Mirach, is second magnitude, and is orange. Above Mirach are two faint stars, and close to the right or upper of these is the Great Spiral. It is near the limit of naked-eye visibility, but there is no mistaking it in binoculars; it shows up as a dim, elongated patch.

We are looking at a whole star-system, larger than our Galaxy. It is over two million light-years away, so we are seeing it as it used to be long before the last Ice Age. Photographs taken with large telescopes are needed to bring out its spiral form, but it is always worth finding even though it looks unspectacular. It is also one of the very nearest of the external galaxies. If we lived on a planet moving round a star in the Spiral, we would see our own Galaxy as a tiny blur.

Two smaller constellations lie below the Andromeda line. *Aries*, the Ram, is in the Zodiac; it is marked by a trio of stars, one of which, Hamal, is second magnitude. Adjoining it is *Triangulum*, the Triangle, which has a characteristic shape even though it is not bright. And below Aries and Triangulum we see *Cetus*, the sea-monster of the Perseus legend (though often referred to simply as the Whale).

Underneath the Square of Pegasus you will find a long, straggly line of faint stars marking another Zodiacal constellation: *Pisces*, the Fishes. Also in the Zodiac, over to the west, are *Aquarius*, the Water-bearer, and *Capricornus*, the Sea-Goat. Neither is of any particular interest, and neither contains any star as bright as second magnitude.

Finally, let us locate the most southerly of all the first-magnitude stars visible from Britain: Fomalhaut, in *Piscis Australis* (the Southern Fish). The Square of Pegasus makes a good guide. The upper right-hand star of the Square, Scheat, is orange; a line passed through the lower right-hand star, Markab, and extended almost down to the horizon will allow you to identify Fomalhaut. It is very low down, and from northern Scotland you will be lucky to see it, but from southern England it is clear enough. Like Vega, it is associated with cool material which may indicate a system of planets. Fomalhaut is a relatively near neighbour; it is only 22 light-years away, and no more than 13 times as luminous as the Sun.

There is only one possible cause of confusion. Over to the left is another solitary star, the second-magnitude Diphda in Cetus, which can be found by using the two left-hand stars of the Square of Pegasus, Alpheratz and Algenib. But Diphda is much higher up than Fomalhaut, and is almost a magnitude fainter.

I know that this survey of the starlit sky has been very incomplete – much has been left out – but at least I hope that it will be enough to help you in identifying the main constellations. If you decide to make a hobby out of astronomy, you will certainly not regret it.

APPENDIX

The leading amateur astronomical society in Britain is the British Astronomical Association (Burlington House, Piccadilly, London W1) which publishes a bi-monthly Journal, and has regular meetings. Many towns and cities have societies of their own; a full list is given in the annual *Yearbook of Astronomy*, published by Macmillan.

Public observatories are unfortunately rare, but some local societies have them. Planetaria include those in London (at the National Maritime Museum) and Chichester (the South Downs Planetarium, opened autumn 2001).

There are many books. Those which I have written include *Atlas of the Universe* (Philip's), *Stargazing* and *Exploring the Night Sky with Binoculars* (both Cambridge University Press), and *The Data Book of Astronomy* (Institute of Physics Publishing). You will find a Philip's Planisphere most helpful. It will show you the sky for any time of the year.

INDEX

ACKNOWLEDGEMENTS

All maps and artworks © Philip's
8 l NASA Goddard Space Flight Center
8 r NASA/JPL/USGS
9 H.J.P. Arnold/Sol Invictus
12 Akira Fujii/David Malin Images
13 Paul Doherty
26 H.J.P. Arnold/Sol Invictus
27 Robin Scagell/Galaxy
37 N.A. Sharp, REU program/NOAO/AURA/NSF
38 Julian Baum/Philip's
39 Paul Doherty
44 Bill Schoening, Vanessa Harvey/REU program/NOAO/AURA/NSF

More titles from the Philip's Astronomy range

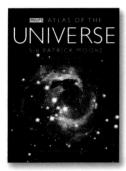

Philip's Atlas of the Universe
Sir Patrick Moore
ISBN 978–0–540–09118–8 £25.00

- Our Solar System and its place in the Universe
- Includes a complete atlas of the constellations and a Moon map

'The best introduction to astronomy'
The Journal of the British Astronomical Association

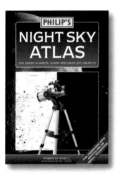

Philip's Night Sky Atlas
Robin Scagell
ISBN 978–0–540–08700–6 £14.99

- Comprehensive star atlas, robustly produced for practical use in the field
- Photo-realistic images to help match the mapping with the sky

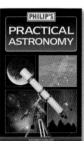

Philip's Practical Astronomy
Storm Dunlop
ISBN 978–0–540–08999–4 £9.99

- Ideal introduction to observational astronomy
- Illustrated with more than 150 colour images

'... attractive, clearly written and includes all that a beginner would need to know to get started' *Popular Astronomy*

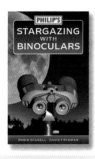

Philip's Stargazing with Binoculars
Robin Scagell & David Frydman
ISBN 978–0–540–09022–8 £7.99

- An essential guide to using binoculars for stargazing
- Includes easy-to-use star maps covering the whole sky

Philip's Planispheres
Rotate the disk to reveal the stars visible from your location at any time on any night of the year. The 51.5°N and 35°S editions come with a season-by-season guide to exploring the skies.

Planisphere 51.5°N
- UK, Northern Europe, Canada
- ISBN 978–0–540–08817–1 £7.99

Planisphere 35°S
- Australia, New Zealand, South America, Southern Africa
- ISBN 978–0–540–08479–1 £10.99

Also available...
Moon Observer's Guide
Peter Grego • ISBN 978–1–84907–065–2 • £9.99

Stargazing with a Telescope
Robin Scagell • ISBN 978–0–540–09023–5 • £7.99

Stargazer
Includes Star Chart, Planisphere and booklet for the Northern Hemisphere.
ISBN 978–0–540–08908–6 • £10.99

Star Finder/Southern Star Finder
Month-by-month calendar for star watchers.
John Woodruff and Wil Tirion
ISBN 978–0–540–08818–8 • £4.99 (N)
ISBN 978–0–540–08093–9 • £5.99 (S)

Star Chart
ISBN 978–1–84907–011–9 • £5.99

Moon Map
ISBN 978–0–540–06378–9 • £6.99

Dark Skies Map
ISBN 978–0–540–08612–2 • £6.99

☎ **How to order** The Philip's range of astronomy titles is available from bookshops or directly from the publisher by phoning 01903 828511 or online at **www.philips-maps.co.uk**
Prices are correct at the time of going to press but are subject to change without notification.